WORKBOOK

Cambridge Assessment
International Education
Endorsed for learner support

Cambridge IGCSE™ and O Level

Additional Mathematics

Val Hanrahan
Jeanette Powell
Series editor: Roger Porkess

HODDER
EDUCATION

Welcome to the *Cambridge IGCSE and O Level Additional Mathematics Workbook*. The aim of this Workbook is to provide you with further opportunity to practise the skills you have acquired while using the *Cambridge IGCSE and O Level Additional Mathematics Student's Book*. It is designed to complement the Student's Book and to provide additional exercises to support you throughout the course and help you to prepare for your examination.

The chapters in this Workbook reflect the topics in the Student's Book. There is no set way to approach using this Workbook. You may wish to use it to supplement your understanding of the different topics as you work through each chapter of the Student's Book, or you may prefer to use it to reinforce your skills in dealing with particular topics as you prepare for your examination. The Workbook is intended to be sufficiently flexible to suit whatever you feel is the best approach for your needs.

All exam-style questions and sample answers in this title were written by the author(s). In examinations, the way marks are awarded may be different.

Every effort has been made to trace all copyright holders, but if any have been inadvertently overlooked, the Publishers will be pleased to make the necessary arrangements at the first opportunity.

Although every effort has been made to ensure that website addresses are correct at time of going to press, Hodder Education cannot be held responsible for the content of any website mentioned in this book. It is sometimes possible to find a relocated web page by typing in the address of the home page for a website in the URL window of your browser.

Hachette UK's policy is to use papers that are natural, renewable and recyclable products and made from wood grown in sustainable forests. The logging and manufacturing processes are expected to conform to the environmental regulations of the country of origin.

Orders: please contact Bookpoint Ltd, 130 Park Drive, Milton Park, Abingdon, Oxon OX14 4SE. Telephone: (44) 01235 827720. Fax: (44) 01235 400401. Email education@bookpoint.co.uk Lines are open from 9 a.m. to 5 p.m., Monday to Saturday, with a 24-hour message answering service. You can also order through our website: www.hoddereducation.com

© Roger Porkess, Val Hanrahan & Jeanette Powell 2018

First published 2018 by

Hodder Education

An Hachette UK Company

Carmelite House

50 Victoria Embankment

London EC4Y 0DZ

www.hoddereducation.com

Impression number   10 9 8 7 6 5 4 3 2 1

Year   2022 2021 2020 2019 2018

Cover photo © Maxal Tamor/Shutterstock

Illustrations by Integra Software Services Pvt. Ltd., Pondicherry, India

Typeset in Times Ten LT Std 11.5/13 by Integra Software Services Pvt. Ltd., Pondicherry, India

Printed in Great Britain by Hobbs the Printers Ltd, Totton, Hampshire, SO40 3WX.

A catalogue record for this title is available from the British Library.

ISBN: 978 1 5104 2165 3

MIX
Paper from
responsible sources
FSC™ C104740

Contents

1 Functions

1 For the function $f(x) = 2x - 5$ find:

   a) $f(2)$ **b)** $f(0)$ **c)** $f(-3)$

2 For the function $g(x) = (2x - 2)^2$ find:

   a) $g(0)$ **b)** $g(0.5)$ **c)** $g(-2)$

3 For the function $h{:}x \rightarrow x^2 - 2x$ find:

   a) $h(2)$ **b)** $h(-2)$ **c)** $h(0)$

4 For the function $f{:}x \rightarrow 5 - 3x$ find:

   a) $f(3)$ **b)** $f(-3)$ **c)** $f(1)$

5 For the function $g(x) = \sqrt{2x + 1}$ find:

   a) $g(0)$ **b)** $g(4)$ **c)** $g(12)$

> Remember that the $\sqrt{}$ symbol means the positive square root of a number.

6 For the function $f(x) = x^2 - 2$
Draw a mapping diagram when the inputs are:

   a) the numbers in the set $\{1, 2, 3\}$ **b)** the numbers in the set $\{-1, -2, -3\}$.

7 Find the range of the following functions:

a) $f(x) = 2x + 1$ Domain $\{0, 1, 2\}$

b) $g(x) = 3x^2 - 1$ Domain $\{2, 4, 6\}$

c) $h(x) = \dfrac{2x - 1}{5}$ Domain $\{1, 3, 6\}$

d) $f(x) = 4x$ Domain $\mathbb{Z}^+$

> The symbol $\mathbb{Z}$ means the set of all integers, and $\mathbb{Z}^+$ means the set of positive integers.

e) $f{:}x \rightarrow x^2 - 5$ Domain $\mathbb{Z}$

8 What values must be excluded from the domain of the following functions and why must they be excluded?

a) $f(x) = \sqrt{2x - 1}$

b) $f(x) = \dfrac{3}{x^2}$

c) $f(x) = \dfrac{2x - 1}{x + 1}$

         Cambridge IGCSE™ and O Level Additional Mathematics Workbook

9 a) On the axes provided, plot the graph of $y = f(x)$, where $f(x) = x^2 - 4$, for $0 \le x \le 4$.

b) Add the line $y = x$ to your graph.

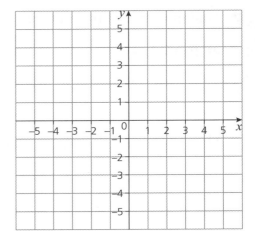

c) Given that $f^{-1}(x) = \sqrt{x+4}$, calculate the values of:

(i) $f^{-1}(-4)$ **(ii)** $f^{-1}(-3)$ **(iii)** $f^{-1}(0)$ **(iv)** $f^{-1}(5)$

d) Use these to add the graph of $y = f^{-1}(x)$ to the graph you drew in part **a)**.

e) What you notice?

Cambridge IGCSE™ and O Level Additional Mathematics Workbook

10 Given that $f(x) = 2x - 3$; $g(x) = x^2$ and $h(x) = 3x - 2$ find the following:

a) $fg(3)$

b) $gf(3)$

c) $fh(2)$

d) $hf(2)$

e) $fgh(0)$

f) $hgf(0)$

g) $fgh(x)$

h) $hgf(x)$

11 Given $f(x) = 2x + 1$ and $g(x) = \sqrt{3x + 1}$ find:

   a) $fg(5)$ **c)** $fg(0)$

   b) $gf(5)$ **d)** $gf(0)$

12 Given $f(x) = x + 3$, $g(x) = x^2 - 3$ and $h(x) = \dfrac{1}{x + 3}$ for $x \neq -3$, find:

   a) $fg(x)$ **d)** $hf(x)$

   b) $gf(x)$ **e)** $hfg(x)$

   c) $fh(x)$ **f)** $hgf(x)$

13 a) Find the inverse for the following functions:

(i) $f(x) = 2x - 3$

(ii) $f(x) = x^2 - 4$ for $x \geqslant 0$

b) On the axes below, plot the graphs of $y = f(x)$ and $y = f^{-1}(x)$.

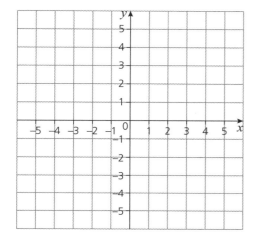

14 The first graph shows the line $y = 2 - x$ and the other graphs are related to this. Write down their equations.

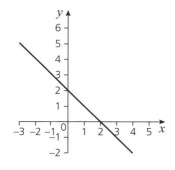

a)

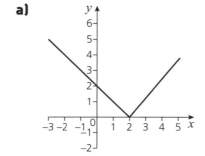

b)

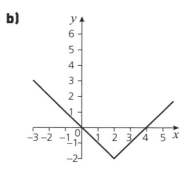

15 The graph below shows part of a quadratic curve, defined for values of $x \geqslant 0$.

a) Determine the equation of the curve.

b) Sketch the inverse of the curve on the same axes and give its equation.

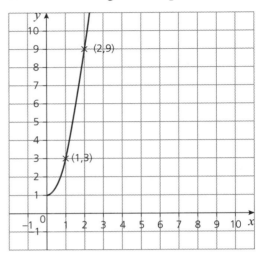

16 Draw the following graphs.

a) $y = x - 2$

b) $y = |x - 2|$

c) $y = -|x - 2|$

d) $y = 2 - |x - 2|$

a)

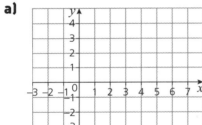

c)

b)

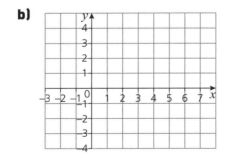

d)

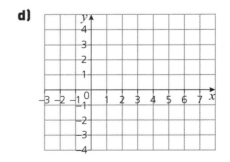

2 Quadratic functions

1 Solve the following equations by factorising:

a) $x^2 + 4x + 4 = 0$

c) $x^2 - 11x - 12 = 0$

b) $x^2 - 3x + 2 = 0$

d) $x^2 + 2x - 15 = 0$

2 Solve the following equations by factorising:

a) $2x^2 + 11x + 12 = 0$

c) $4x^2 - 12x + 9 = 0$

b) $3x^2 - 17x - 6 = 0$

d) $4x^2 + 5x - 6 = 0$

3 Solve the following equations:

a) $x^2 - 64 = 0$

c) $49 - 16x^2 = 0$

b) $9x^2 - 144 = 0$

d) $64x^2 - 100 = 0$

4 For each of the following functions:

   (i) factorise the function

   (ii) work out the coordinates of the stationary point

   (iii) state whether the stationary point is a maximum or a minimum.

   a) $y = x^2 + 7x + 12$

   c) $f(x) = 2x^2 + x - 28$

   b) $y = 12 + 2x - 2x^2$

   d) $f(x) = 6 + 2x - 8x^2$

5 For each of the following

   (i) write the left hand side in the form $c(x + a)^2 + b$

   (ii) solve the equation.

   a) $2x^2 - 10x + 15 = 0$

   c) $2x^2 + 8x - 8 = 0$

   b) $3x^2 - 6x + 10 = 0$

   d) $5x^2 + 15x + 9 = 0$

6 For each of the following functions:

    (i) use the method of completing the square to find the coordinates of the stationary point

    (ii) state whether the stationary point is a maximum or a minimum.

a) $y = x^2 + 2x - 12$ **c)** $f(x) = 15 + 2x - x^2$

b) $f(x) = x^2 + 3x + 9$ **d)** $y = 3 + 4x - 2x^2$

7 Draw the graph and find the corresponding range for each function and domain.

a) $y = x^2 - 2x - 8$ for the   **b)** $f(x) = 4x^2 - 2x - 12$ for the
    domain $-3 \leqslant x \leqslant 5$ domain $-3 \leqslant x \leqslant 3$

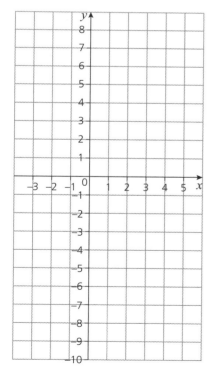

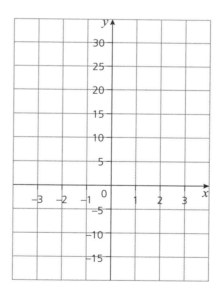

8 For each of the following equations, use the discriminant to decide
if there are two real and different roots, two equal roots or no real roots.
Solve the equations with real roots.

a) $3x^2 - 6x = 0$

c) $r^2 + 5r - 14 = 0$

b) $m^2 + 6m + 9 = 0$

d) $2x^2 - 7x + 6 = 0$

9 Solve the following equations by

(i) completing the square

(ii) using the quadratic formula.

a) $x^2 - 4x - 9 = 0$

c) $2r^2 + 2r - 1 = 0$

(i)

(i)

(ii)

(ii)

b) $y^2 + 3y = 5$

d) $3m^2 - 12m + 7 = 0$

(i)

(i)

(ii)

(ii)

10 For each pair of equations:

    (i) determine if the line intersects the curve, is a tangent to the curve or does not meet the curve

    (ii) give the coordinates of any points where the curve and line touch or intersect.

a) $y = x^2 + 2x - 3; y = x - 1$

c) $y = 2 - x - x^2; y = 2 - x$

b) $y = x^2 - 3x - 3; y = x - 8$

d) $y = x^2 + 2x - 5; y = 4 - 2x$

11 Solve the following inequalities and illustrate each solution on a number line:

a) $x^2 - 5x + 6 > 0$

b) $p^2 + 3p - 10 < 0$

c) $4 \geqslant m^2 + 3m$

Cambridge IGCSE™ and O Level Additional Mathematics Workbook

3 Equations, inequalities and graphs

1 a) Plot the graphs of $y = 3x$ and $y = |3x|$ on the same axes.

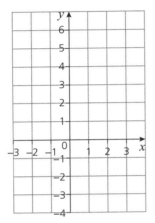

b) Plot the graphs of $y = x - 4$ and $y = |x - 4|$ on the same axes.

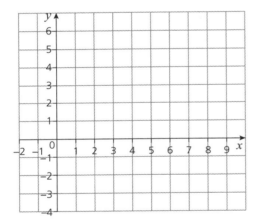

c) Plot the graphs of $y = 2x - 4$ and $y = |2x - 4|$ on the same axes.

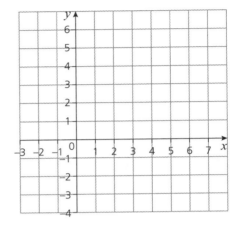

d) Plot the graphs of $y = 5 - x$ and $y = |5 - x|$ on the same axes.

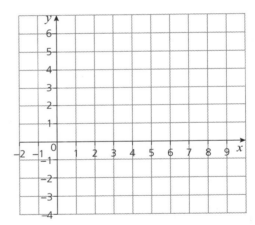

   Cambridge IGCSE™ and O Level Additional Mathematics Workbook

2 a) On the axes below, draw the graph of $y = |x + 2|$.

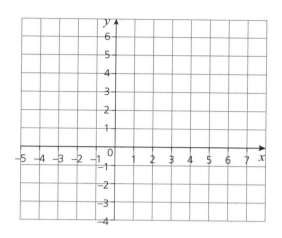

b) Use the graph to solve $|x + 2| = 2$.

c) Use algebra to verify your answer to **b)**.

3 a) On the axes below, draw the graph of $y = |2x - 3|$.

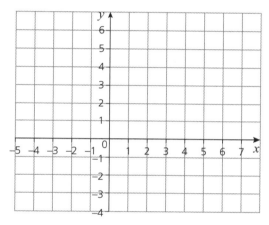

b) Use the graph to solve $|2x - 3| = 1$.

c) Use algebra to verify your answer to **b)**.

4 Solve the equation $|x - 2| = |x + 2|$

a) graphically

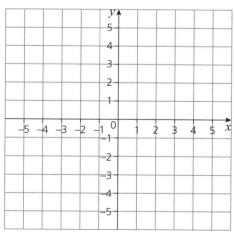

b) algebraically.

5 Solve the equation $|2x + 3| = |2x - 3|$

a) graphically

b) algebraically.

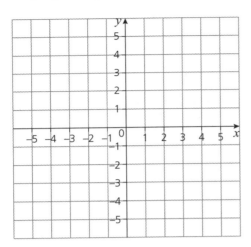

6 Write each of the following inequalities in the form $|x - a| \leqslant b$:

a) $-2 \leqslant x \leqslant 12$

b) $-5 \leqslant x \leqslant 25$

c) $-16 \leqslant x \leqslant 8$.

7 Write each of the following expressions in the form $a \leqslant x \leqslant b$.

a) $|x + 1| \leqslant 3$

b) $|x + 2| \leqslant 4$

c) $|x + 3| \leqslant 5$

8 **(i)** Solve the following inequalities and   **(ii)** illustrate the solution on a number line:

a) $|x + 1| < 5$

b) $|x + 1| > 5$

c) $|3x + 2| \leqslant 7$

d) $|3x + 2| \geqslant 7$

9 Illustrate the following inequalities by shading out the unwanted region:

a) $y + 2x < 0$

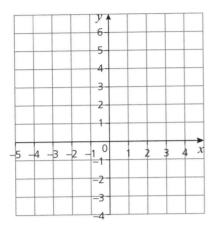

c) $2y - 3x < 0$

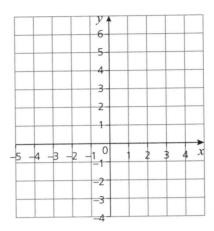

b) $y - 2x > 0$

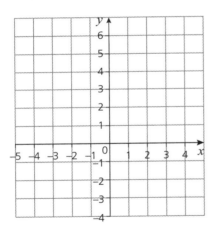

d) $2y + 3x > 0$

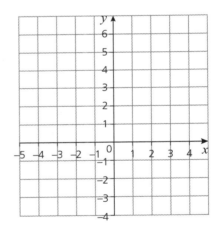

10 a) Draw the lines $y = x - 3$ and $y = x + 3$ on the grid

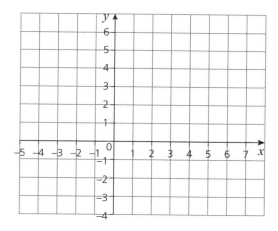

b) Hence solve these inequalities

(i) $|x - 3| < |x + 3|$

(ii) $|x - 3| > |x + 3|$

11 Solve the following inequalities algebraically:

a) $|2x - 3| < |x + 3|$

b) $|2x - 3| > |x + 3|$

12 The unshaded region of each graph illustrates an inequality of a modulus function. In each case write the inequality.

a) $y = |2x - 1|$

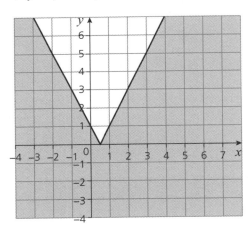

b) $y = |3 - 2x|$

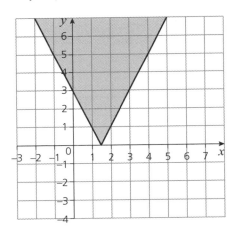

13 Sketch the following graphs on the axes provided, indicating the points where they cross the co-ordinate axes:

a) $y = x(x + 1)(x + 2)$

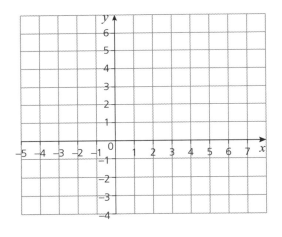

b) $y = |x(x + 1)(x + 2)|$

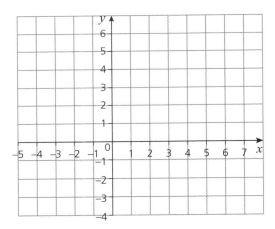

                                      Cambridge IGCSE™ and O Level Additional Mathematics Workbook

14 Identify the following cubic equations from their graphs:

a)

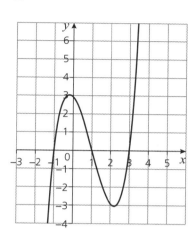

b)

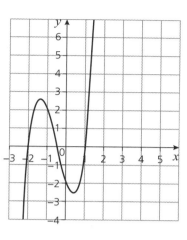

c)

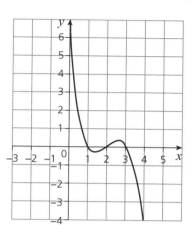

15 Find an equation for each of the following modulus graphs.
All represent the moduli of cubic graphs.

a)

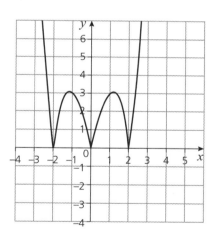

b)

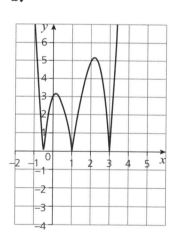

c)

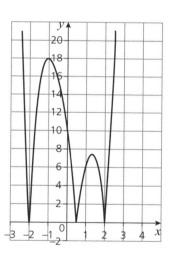

4 Indices and surds

1 Simplify the following; give your answers in the form x^n.

a) $3^4 \times 3^8$

c) $5^4 \div 5^3$

e) $(2^4)^2$

b) $4^{-2} \times 4^7$

d) $6^5 \div 6^{-2}$

f) $(7^2)^{-3}$

2 Simplify the following; leave your answers in standard form.

a) $(5 \times 10^4) \times (3 \times 10^3)$

c) $(8 \times 10^4) \div (2 \times 10^2)$

b) $(2 \times 10^6) \times (4 \times 10^{-3})$

d) $(6 \times 10^9) \div (2 \times 10^{-3})$

3 Rewrite each of the following as a number raised to a positive integer power.

a) 2^{-4}

c) $\left(\frac{1}{2}\right)^{-2}$

b) 4^{-3}

d) $\left(\frac{2}{5}\right)^{-3}$

Cambridge IGCSE™ and O Level Additional Mathematics Workbook

4 Find the value of each of the following. Answer as a whole number or fraction.

   a) $5^3 \times 5^{-2}$ **d)** $(3^2)^3$ **g)** $16^{-\frac{1}{2}}$

   b) $7^{-5} \times 7^3$ **e)** $\left(\dfrac{2}{3}\right)^{-2}$ **h)** $25^{\frac{3}{2}}$

   c) $2^8 \div 2^3$ **f)** $\left(\dfrac{4}{5}\right)^3$ **i)** $19^7 \times 19^{-7}$

5 Rank each set of numbers in order of increasing size.

   a) $3^3, 4^2, 2^5$ **b)** $4^{-3}, 7^{-2}, 3^{-4}$

6 Find the value of x.

   a) $\dfrac{2^x \div 2^2}{2^4 \times 2^3} = 2^4$ **b)** $\dfrac{(7^5 \times 7)^x}{7^3 \times 7^2} = 7^7$

7 Simplify:

   a) $3n^3 \times 2n^2$ **d)** $9b^5 \div 3b^2$

   b) $5p^2q^4 \times 3q^{-2}$ **e)** $6x^2y^{-5} \times 4x^6y^{-4}$

   c) $(2a^3)^3$ **f)** $\dfrac{20a^2b^3}{4a^4b^5}$

8 Find integers a and b such that

a) $5^a \times 2^b = 10^3$

b) $10^a \div 5^b = 2^6 \times 5^2$.

$a =$ $b =$

$a =$ $b =$

9 Write each of the following in its simplest form.

a) $\sqrt{72}$

c) $\sqrt{32} + \sqrt{128}$

b) $2\sqrt{2} + 5\sqrt{2}$

d) $3\sqrt{112} - 2\sqrt{28}$

10 Express each of the following as the square root of a single number:

a) $2\sqrt{5}$

c) $9\sqrt{2}$

b) $3\sqrt{7}$

d) $6\sqrt{3}$

11 Write each of the following in the form $a\sqrt{b}$ where a and b are integers and b is as small as possible.

> A rational number is an integer or fraction.

a) $\dfrac{\sqrt{72}}{64}$

c) $\dfrac{\sqrt{18}}{8}$

b) $\dfrac{\sqrt{75}}{16}$

d) $\dfrac{\sqrt{32}}{144}$

12 Simplify the following by collecting like terms:

a) $(2 + \sqrt{3}) + (6 + 3\sqrt{3})$

b) $5(\sqrt{2} + 1) - 3(1 - \sqrt{2})$

13 Expand and simplify:

a) $(5 + \sqrt{5})(5 - \sqrt{5})$

c) $(6 + \sqrt{3})^2$

b) $\sqrt{2}(6 + \sqrt{2})$

d) $(5 - 2\sqrt{2})(5 + 2\sqrt{2})$

14 Rationalise the denominators. Give each answer in its simplest form.

a) $\dfrac{1}{\sqrt{5}}$

c) $\dfrac{1}{\sqrt{6} - 2}$

b) $\dfrac{3}{\sqrt{3}}$

d) $\dfrac{5 + \sqrt{2}}{3 - \sqrt{2}}$

15 Write the following in the form $a + b\sqrt{c}$ where c is an integer and a and b are rational numbers.

a) $\dfrac{7 + \sqrt{3}}{2 - \sqrt{3}}$

b) $\dfrac{3\sqrt{3}}{\sqrt{3} - 1}$

16 Work out the length of AB. Answer in the simplest surd form.

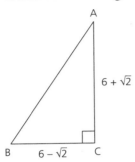

AB =

17 A rectangle has sides of length x cm and $2x$ cm and a diagonal of length 15 cm.

a) Use Pythagoras' Theorem to find the exact value of x in its simplest surd form.

b) Work out the area of the rectangle.

18 An isosceles triangle has sides of length $\sqrt{5}$ cm, $\sqrt{5}$ cm and $\sqrt{2}$ cm. Work out:

a) the height of the triangle in its simplest surd form

b) the area of the triangle.

5 Factors of polynomials

1 Multiply $(x^3 + 2x^2 + x - 4)$ by $(x + 2)$.

2 Multiply $(2x^3 - 3x^2 + 2x + 2)$ by $(x - 1)$.

3 Multiply $(2x^3 - 5x^2 + 4)$ by $(2x - 1)$.

4 Multiply $(x^3 + 3x^2 - 5x + 5)$ by $(x + 3)$.

5 Multiply $(2x^2 - 5x + 6)$ by $(x^2 + x - 2)$.

6 Divide $(x^3 + x^2 - 4x - 4)$ by $(x + 1)$.

7 Divide $(x^3 + 7x^2 + 16x + 12)$ by $(x + 2)$.

8 Divide $(x^3 - 13x + 12)$ by $(x - 1)$.

9 Simplify $(2x^2 - 7x + 1)^2$.

10 Determine whether the following linear functions are factors of the given polynomials or not.

a) $(x^3 + 9x^2 - 2x + 4); (x - 1)$

b) $(x^3 - x^2 - x + 1); (x + 1)$

c) $(2x^3 - 2x^2 + 5x - 5); (x - 1)$

11 For each equation:

    (i)   use the factor theorem to find a factor of each function

    (ii) factorise each function as a product of three linear factors

    (iii) draw its graph on the axes provided.

a) $x^3 - 3x^2 - x + 3$

    (i)

    (iii)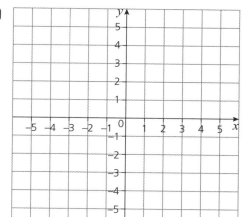

    (ii)

b) $x^3 + 2x^2 - 5x - 6$

    (i)

    (iii)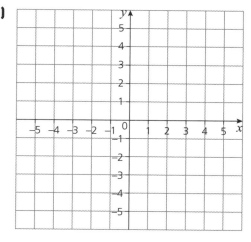

    (ii)

Cambridge IGCSE™ and O Level Additional Mathematics Workbook

12 Factorise as far as possible: $x^3 + 6x^2 + 12x + 7$.

13 For what value of a is $(x - 3)$ a factor of $x^3 - ax^2 + 18$?

14 For what value of b is $(2x + 1)$ a factor of $2x^3 - 7x^2 - bx - 6$?

15 The expression $x^3 + px^2 - 6x + q$ is exactly divisible by $(x + 1)$ and $(x + 4)$.

Find and solve two simultaneous equations to find p and q.

16 Find the remainder when each function is divided by the linear factor (shown in brackets).

a) $x^3 + 3x^2 - 2x + 1; (x - 2)$

b) $3x^3 + 3x^2 - 4x - 14; (x - 1)$

17 The equation $f(x) = x^3 + 3x^2 - 10x - 24$ has three integer roots. Solve $f(x) = 0$.

18 When $x^3 + ax^2 + bx + 2$ is divided by $(x - 2)$ the remainder is 36.
When it is divided by $(x + 3)$ the remainder is −4.

a) Find the values of a and b.

b) Solve the equation $x^3 + ax^2 + bx + 2 = 0$.

6 Simultaneous equations

1 Solve the following pairs of simultaneous equations graphically.

a) $y = x + 4$

$y = 2x + 2$

b) $x + 3y = 9$

$2x - y = 4$

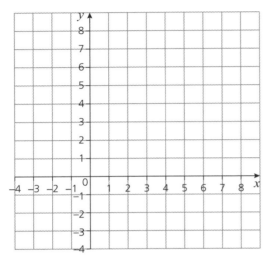

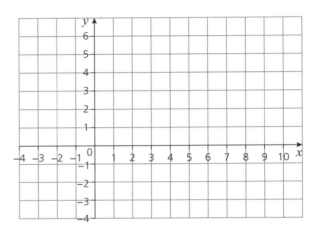

Use the substitution method to solve the simultaneous equations in questions **2** and **3**.

2 $2x + y = 13$

$y = 2x + 1$

3 $3x + 2y = 6$

$y = x - 2$

Use the elimination method to solve the simultaneous equations in questions **4** and **5**.

4 $x + y = 6$

$x - y = 2$

5 $2x + y = 10$

$3x - y = 5$

6 Four tins of soup and 3 packs of bread rolls cost $10:80.
Two tins of soup and 5 packs of bread rolls cost $9:60.
Find the cost of 3 tins of soup and 7 packs of bread rolls.

7 In a sale, 3 DVDs and 4 CDs cost $51 and 4 DVDs and 3 CDs cost $54.

a) Find the cost of a DVD and the cost of a CD.

b) I have $90 dollars to spend and would like to buy the same number of DVDs and CDs. How many of each can I buy?

8 a) Solve this pair of simultaneous equations algebraically:

$$x^2 - 6x - y = -8$$

$$y - x + 4 = 0$$

b) Illustrate the solution graphically.

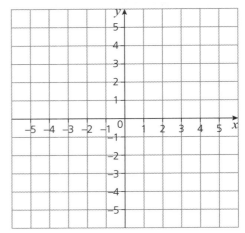

9 Solve this pair of simultaneous equations algebraically:

$$x^2 - y^2 + xy = 20$$

$$x = 2y$$

10 Two numbers, x and y, have a difference of 2 and a product of 15.

  a) Write down two equations that are satisfied by x and y.

  b) Find two possible values for the pair of numbers by solving the equations simultaneously.

      Cambridge IGCSE™ and O Level Additional Mathematics Workbook

11 a) Solve this pair of simultaneous equations algebraically:

$$(x-2)^2 + (y-3)^2 = 9$$

$$x - y + 4 = 0$$

b) Given that $(x-2)^2 + (y-3)^2 = 9$ is the equation of a circle, with centre (2,3) and radius $\sqrt{9}$, illustrate the solution graphically.

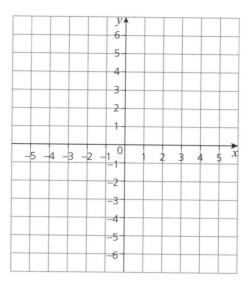

7 Logarithmic and exponential functions

> Remember that, for example, the relationship $2^3 = 8$ can be written as the equation $\log_2 8 = 3$.

1 In each part of this question, find the values of x and y.

    a) $4^x = 16, y = \log_4 16$ **b)** $x^4 = 81, y = \log_3 81$ **c)** $x = 3^{-2}, y = \log_3 x$

2 Using your knowledge of indices, and without using your calculator, find the values of:

    a) $\log_2 32$ **b)** $\log_3 9$ **c)** $\log_5 \dfrac{1}{25}$

3 Use the rules for manipulating logarithms to write each of the following as a single logarithm.

    a) $\log 6 + \log 2$ **d)** $\dfrac{1}{2}\log 9 - \log 3$

> For example, $\log 4 + \log 3$ can be written as $\log(4 \times 3) = \log 12$

    b) $\log 36 - \log 9$ **e)** $\dfrac{1}{2}\log 25 + \dfrac{1}{3}\log 27$

    c) $5\log 2$ **f)** $\log 5 + \log 4 - \log 2$

      Cambridge IGCSE™ and O Level Additional Mathematics Workbook

4 Express each of the following in terms of $\log x$.

a) $\log x^3 - \log x^2$

c) $2\log\sqrt{x} + \log x$

b) $2\log x^3 + 3\log x^2$

d) $6\log^3\sqrt{x} - 4\log\sqrt{x}$

5 a) Express $\log_5 \dfrac{1}{x^2} + \log_5 x^4$ as a multiple of $\log_5 x$.

b) Hence solve the equation $\log_5 \dfrac{1}{x^2} + \log_5 x^4 = 2$ without using your calculator.

6 Draw each of the following graphs on the axes below.
In each case show the vertical asymptote and the coordinates
of the points where the graph crosses the x axes.

a) $y = \log_{10} x$ **b)** $y = \log_{10}(x + 5)$ **c)** $y = \log_{10}(x - 10)$

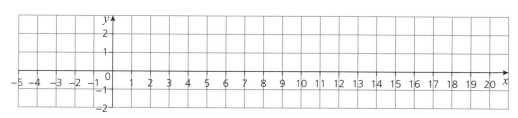

7 For each of the following graphs:

a) Starting with the curve $y = \lg x$ state the transformation (in order when more than one is needed) required to sketch the curve:

> Notice that, as in this question, $\log_{10} x$ is often written as $\lg x$.

(i) $y = 2\lg x$ **(ii)** $y = \lg 2x$.

b) Sketch the curves on the axes below, together with the curve $y = \lg x$ in each case.

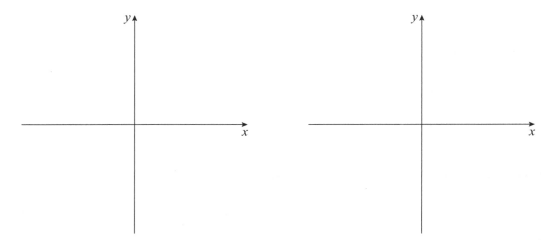

c) Would the results be the same, or different, if logarithms to a different base were used?

8 Match the correct equation to each graph.

Equation	Graph	Equation	Graph
$y = \ln(x + 2)$		$y = \ln x + 2$	
$y = \ln 2x$		$y = \ln(2 - x)$	
$y = \ln(x - 2)$		$y = \ln x - 2$	

Notice that, as in this question, $\log_e x$ is usually written as $\ln x$.

a)

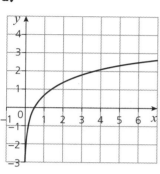

c)

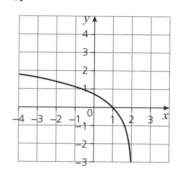

e)

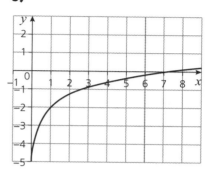

b)

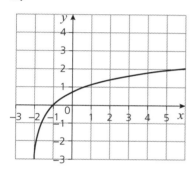

d)

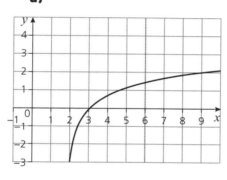

f)

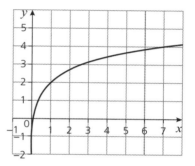

9 Solve the following equations for x, given that $\lg p = 5$.

a) $p = 10^x$

b) $p^{2x} - 6p^x + 8 = 0$

10 Use logarithms to solve the equation $3^{2x-1} = 2^{3x+1}$ giving your answer to 3 s.f.

11 The formula for compound interest is $A = P\left(1 + \dfrac{R}{100}\right)^n$ where A represents the final amount, P the principal (the amount invested), R the rate of interest and n the number of years.

   a) For how long, to the nearest month, would \$10 000 need to be invested to produce a final amount of \$15 000 if the rate of interest is 2.8%?

   b) \$10 000 is invested for 5 years. What was the rate of interest if the final amount is \$12 000? Answer to 1 decimal place.

12 Sketch the curves of the given functions on the grids below.
Show any asymptotes and give the coordinates of any points of intersection with the axes.

a) $y = e^x$, $y = e^x - 1$ and $y = e^{x-1}$

b) $y = e^x$, $y = 3e^x$ and $y = e^{3x}$

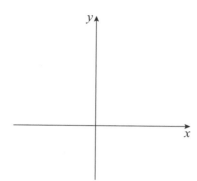

13 Sketch each of the following curves. In each case write the equation of the asymptote and the coordinates of the point where they cross the y-axis.

a) $y = e^x + 1$

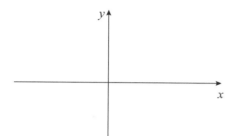

b) $y = e^{-x} + 1$

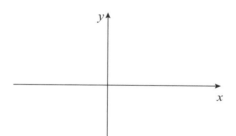

14 Solve the following equations giving your answers to 3 s.f.

a) $4e^{5t} = 30$

c) $e^{t+3} = 30$

b) $5e^{4t} = 30$

d) $e^{t-3} = 30$

15 Match the correct equation to each of the graphs.

Equation	Graph
$y = e^{x-1}$	
$y = e^{x+1}$	
$y = e^x + 1$	
$y = e^x - 1$	

a)

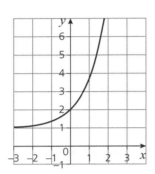

c)

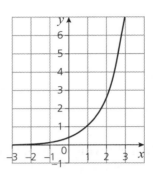

b)

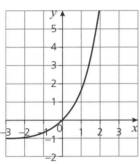

d)

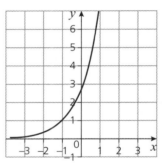

16 In general, continuous compound interest on an investment is given by the formula $A = Pe^{\frac{r}{100} \times t}$, where \$$P$ is the amount invested for t years at a rate of r%, giving a final amount of \$$A$.

  a) To the nearest \$, how much will an amount of \$20 000 invested at a rate of 4% be worth in 5 years' time?

  b) How long would \$20 000 need to be invested at 4% to be worth \$30 000?
  Give your answer to the nearest month.

17 A radioactive substance of mass 250 g is decaying so that the mass M left after t days is given by the formula $M = 250e^{-0.0035t}$.

a) On the axes below, sketch the graph of M against t.

b) How much, to the nearest gram, is left after one year (i.e. 365 days)?

c) What is the rate of decay after 365 days?

d) After how long, to the nearest day, will there be less than 10 g remaining?

Straight line graphs

1   A(2, 3), B(6, 4) and C(5, 0) are the vertices of a triangle.

a)   Draw the triangle ABC on the axes provided.

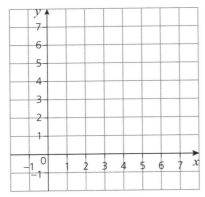

b)   Show by calculation that ABC is an isosceles triangle and write down the two equal sides.

c)   Find the coordinates of D, the midpoint of AC.

d)   Prove that BD is perpendicular to AC.

e)   Find the area of the triangle.

2 The graph shows a quadrilateral PQRS.

   a) Prove, using suitable calculations, that PQRS is a trapezium.

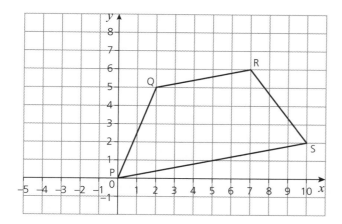

   b) PQRT is a parallelogram. Find the coordinates of T.

   c) Prove that PQRT is not a rhombus.

3 In each of parts (a) and (b) you are given the equation of a line and the coordinates of a point. Find the equation of the line through the given point that is

   (i) parallel to the given line   **(ii)** perpendicular to the given line.

     a) $y = 3x - 2$; $(3,1)$ **b)** $2x + y = 2$; $(-1,-2)$

4 Find the equation of the perpendicular bisector of the line joining each pair of points.

a) $(1, 4)$ and $(5, 2)$

b) $(-3, -4)$ and $(2, 4)$

5 A median of a triangle is a line joining one of the vertices to the midpoint of the opposite side. In a triangle ABC, A is the point $(0, 6)$, B is $(4, 8)$ and C is the point $(2, -2)$.

a) Sketch the triangle on the axes provided.

b) Find the equations of the three medians of the triangle.

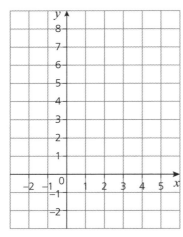

c) Show that the three medians are concurrent (i.e. all three intersect at the same point).

6 Match the equivalent relationships:

a) $\log y = \log b + x \log a$

c) $\log y = \log x + b \log a$

b) $\log y = \log a + b \log x$

d) $\log y = \log a + x \log b$

	Matches to
$y = ax^b$	
$y = xa^b$	
$y = ab^x$	
$y = ba^x$	

The relationship $p = qr^n$ can be written using logarithms as $p = \log q + \log r^n$ and so is equivalent to $\log p = \log q + n \log r$.

7 In this question k and a are constants. In each of the following cases:

i) write the equation with $\ln y$ as subject;

ii) given that $\ln y$ is on the vertical axis and the graph is a straight line, state the quantity on the horizontal axis;

iii) state the gradient of the line and its intercept with the vertical axis.

a) $y = ax^k$

i)

ii)

iii)

b) $y = kx^a$

i)

ii)

iii)

c) $y = ka^x$

i)

ii)

iii)

d) $y = ak^x$

i)

ii)

iii)

8   The results of an experiment are shown in the table:

p	2	4	6	8	10	12
q	7.1	10.0	12.2	14.1	15.8	17.3

The relationship between the two variables, p and q, is of the form $q = Ap^b$ where A and b are constants.

a) Show that the relationship may be written as $\lg q = b \lg p + \lg A$.

b) What graph must be plotted to test this model?

c) Plot the graph on the axes provided and use it to estimate the values of b and A.

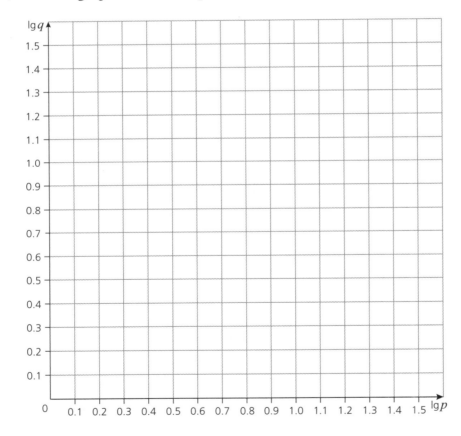

9 Circular measure

1 Express each angle in radians. Leave your answer in terms of π.

> 0.3^c is another way of writing 0.3 radians.

a) 60°

c) 27°

b) 270°

d) 108°

2 Express each angle in degrees. Answer to 1 d.p. where necessary.

a) $\dfrac{\pi}{3}$

c) 0.3^c

b) $\dfrac{2\pi}{9}$

d) $\dfrac{3\pi}{5}$

3 Complete the table, which gives information about some sectors of circles.

Radius, r (cm)	Angle at centre, θ (degrees)	Arc length, s (cm)	Area, A (cm^2)
12	150		
8		20	
	75	12	
15			100
	30		60

4 Complete the table, which gives information about some sectors of circles. Leave your answers as a multiple of π where possible.

Radius, r (cm)	Angle at centre θ (radians)	Arc length, s (cm)	Area A (cm^2)
8	$\dfrac{2\pi}{3}$		
15		15	
	$\dfrac{\pi}{4}$	12	
6			20π
	$\dfrac{2\pi}{5}$		50

5   Look at this diagram:

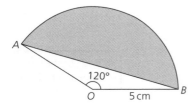

a) Calculate the area of the sector OAB.

b) Calculate the area of the triangle OAB.

c) Find the area of the shaded segment.

6   The diagram shows two circles, each of radius 5 cm, with each one passing through the centre of the other.

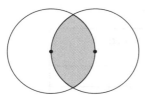

a) Calculate the area of the shaded region.

b) Calculate the perimeter of the shaded region.

7 The shaded region in the diagram is the top of a desk that is to be covered in leather. AB and DC are arcs of circles with centre O and radii and angle as shown.

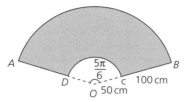

a) Work out the area of the desk to be covered. Give the answer in square metres.

b) **(i)** The leather is sold in rectangular strips 140 cm wide, and is sold in units of 10 cm. What length must be purchased?

(ii) How much is wasted?

10 Trigonometry

1 Find the length of x in each triangle. Give your answer to 2 d.p.

a)

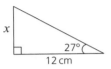

b)

c)

2 Write the following in terms of a single trigonometric function.

a) $\dfrac{\sin\theta}{\cos\theta}$

b) $\dfrac{\sin\theta}{\tan\theta}$

c) $\cos\theta \times \tan\theta$

3 Simplify:

a) $\cos^2\theta(1 + \tan^2\theta)$

b) $\tan^2\theta(1 - \sin^2\theta)$

 Cambridge IGCSE™ and O Level Additional Mathematics Workbook

4 In the diagram, OA = 1 cm, angle
AOB = angle BOC = angle COD = 60° and
angle OAB = angle OBC = angle OCD = 90°.

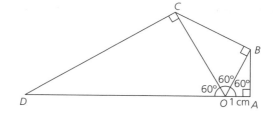

a) Find the length of OD.

b) Show that the perimeter of OABCD is $(9 + 7\sqrt{3})$ cm.

5 Work out the values of the following quantities without using a calculator.
Show your working carefully.

a) $\sin^2 30° - \cos^2 30° \tan^2 30°$

b) $\sin^2 \dfrac{\pi}{4} - \cos^2 \dfrac{\pi}{4} \tan^2 \dfrac{\pi}{4}$

c) $\sin^2 60° - \cos^2 60° \tan^2 60°$

6 a) By plotting suitable points, draw the curve of $y = \cos x$ for $0° \leqslant x \leqslant 360°$ on the grid below.

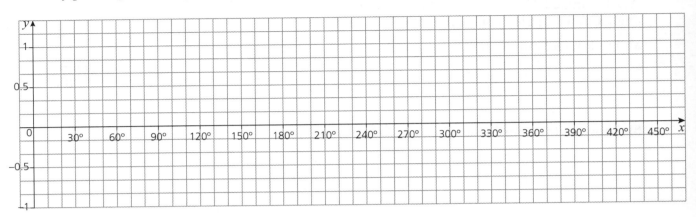

b) Solve the equation $\cos x = 0.4$ for $0° \leqslant x \leqslant 360°$ and illustrate the roots on your sketch.

c) Write down, without using your calculator, the solution to the equation $\cos x = -0.4$ for $0° \leqslant x \leqslant 360°$.

7 Without using your calculator, write the following as fractions or using surds.

a) $\sin 60°$

b) $\cos 120°$

c) $\tan 150°$

8 Solve the following equations for $0 \leqslant x \leqslant 2\pi$ without using your calculator.

a) $\sin \theta = \dfrac{\sqrt{3}}{2}$

b) $\cos \theta = \dfrac{\sqrt{3}}{2}$

c) $\tan \theta = -1$

9 Without using a calculator show that:

a) $\sin^2 30° + \sin^2 45° = \sin^2 60°$

b) $3\cos^2 \dfrac{\pi}{3} = \sin^2 \dfrac{\pi}{3}$

10 Solve the following equations for $-360° \leqslant x \leqslant 360°$.

a) $\sin(x - 30°) = 0.6$

b) $\cos(x + 60°) = 0.4$

c) $\tan(x - 45°) = 1$

11 Starting with the graph of $y = \sin x$, state the transformations that can be used to sketch each of the following curves.

a) $y = \sin 3x$

b) $y = 2\sin 3x$

c) $y = 2\sin 3x - 1$

12 Apply these transformations to the graph of $y = \sin x$. State the equation, amplitude and period of each transformed graph.

a) A stretch of scale factor $\frac{1}{2}$ parallel to the x-axis.

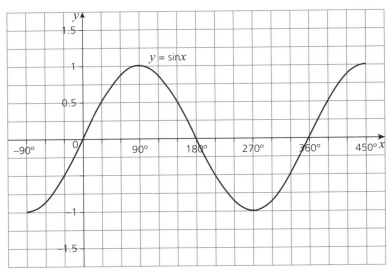

b) A translation of 90° in the negative x direction.

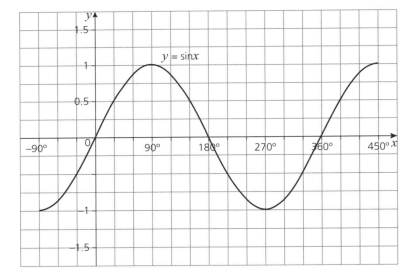

c) A stretch of scale factor 2 parallel to the *y*-axis followed by a translation of 1 unit vertically downwards.

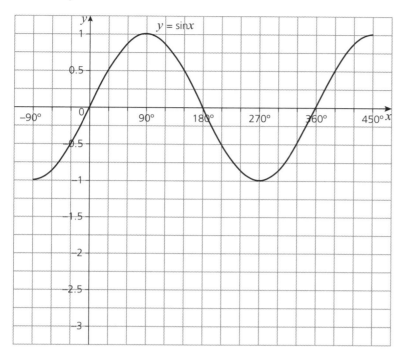

d) A translation of 1 unit vertically downwards followed by a stretch of scale factor 2 parallel to the *y*-axis.

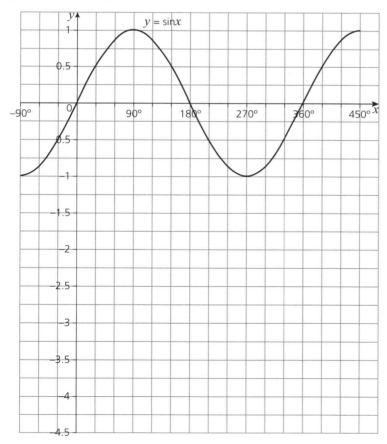

13 State the transformations required, in the correct order, to obtain the graph below from the graph of $y = \sin x$.

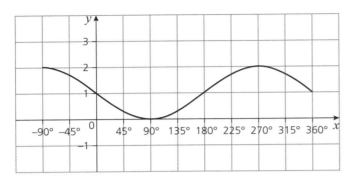

14 State the transformations required, in the correct order, to obtain the graph below from the graph of $y = \tan x$.

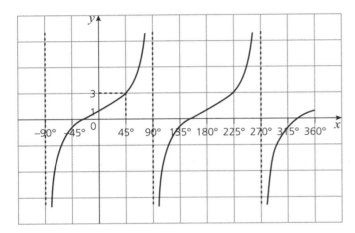

15 State the transformations required, in the correct order, to obtain the graph below from the graph of $y = \cos x$.

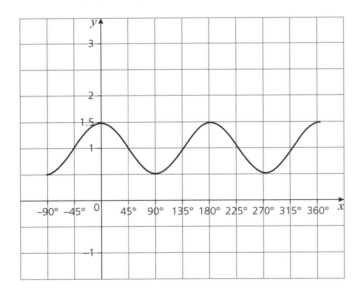

16 Simplify:

a) $\dfrac{\sin\theta}{\cos\theta} + \dfrac{\cos\theta}{\sin\theta}$

c) $\dfrac{1}{\cos\theta\sqrt{(1+\tan^2\theta)}}$

b) $\dfrac{\sqrt{(1+\tan^2\theta)}}{\sqrt{(1-\sin^2\theta)}}$

d) $\dfrac{1-\sec^2\theta}{1-\operatorname{cosec}^2\theta}$

17 Solve $\cot x = \sin x$ for $0° \leqslant x \leqslant 360°$.

18 Solve $\tan x + \cot x = 2\sec x$ for $0 \leqslant x \leqslant 2\pi$.

11 Permutations and combinations

1 Without using a calculator, evaluate the following.

a) $6!$

b) $\dfrac{8!}{6!}$

c) $\dfrac{5! \times 8!}{6! \times 3!}$

2 Simplify the following.

a) $\dfrac{(n+2)!}{n!}$

b) $\dfrac{(n-2)!}{n!}$

c) $\dfrac{(2n+1)!}{(2n-1)!}$

3 Factorise:

a) $7! - 5!$

b) $(n+1)! - (n-1)!$

4 How many different five letter arrangements can be formed from the letters A, B, C, D and E if letters cannot be repeated?

5 Eight friends are going to the theatre together and they all have tickets for adjacent seats in the same row. In how many ways can they be seated?

6 How many different arrangements are there of the letters in each word?

a) CHINA

b) ISLAND

c) DAUGHTER

d) UNIVERSAL

7 A security keypad has the numbers 1, 2, 3, 4 and the letters A, B, C, D on it.
In order to unlock it, a passcode with five of these numbers and letters is needed.
How many possible passcodes are there if:

a) there are no restrictions

b) there must be at least two letters and at least two numbers?

8 Without using a calculator, evaluate the following.

a) $^{10}P_2$

b) $^8P_2$

c) $^6P_2$

d) $^4P_2$

9 Without using a calculator, evaluate the following.

a) $^{10}C_8$

b) $^8C_6$

c) $^6C_4$

d) $^4C_2$

10 There are ten cleaners at a supermarket, one of whom is supervisor. If only 6 cleaners are needed for each shift, including the supervisor, how many possible teams are there?

11 A quiz team of 4 people is chosen at random from 5 girls and 7 boys.
In how many ways can the team be chosen if:

a) there are no restrictions

b) there must be equal numbers of boys and girls

c) there must be more boys than girls?

12 Four letters are picked from the word MAJESTIC. In how many of these choices is there at least one of the letters A, E or I among the letters?

12 Series

The word 'expand' means 'write out term by term'. So expanding $(x + 1)^2$ gives $x^2 + 2x + 1$.

In the questions on this page, simplify the terms in your expansions as far as possible.

1 Expand the following binomial expressions:

a) $(1 + x)^5$ **b)** $(1 - x)^5$ **c)** $(1 + 2x)^5$

2 Expand the following binomial expressions:

a) $(2x + y)^3$ **b)** $(2x - y)^3$ **c)** $(2x + 3y)^3$

3 Find the first three terms, in ascending powers of x, in the expansions of:

a) $(3 - x)^5$ **b)** $\left(3 - \dfrac{x}{2}\right)^5$

4 Find the first three terms, in descending powers of x, in the expansion of the following:

a) $\left(2 - \dfrac{1}{x}\right)^4$ **b)** $\left(3 - \dfrac{2}{x}\right)^4$

      Cambridge IGCSE™ and O Level Additional Mathematics Workbook

5 Find:

a) the coefficient of x^2 in the expansion of $(1 + 2x)^6$

b) the coefficient of x^3 in the expansion of $(1 + 2x)^7$.

6 a) Expand $(1 - 2x)^4$.

b) Hence expand $(1 + x)(1 - 2x)^4$.

7 Identify which of the following sequences are arithmetic, stating the common difference where appropriate.

	Sequence	Arithmetic? Yes / No	Common difference
a)	$1, 5, 9, 13, \ldots$		
b)	$2, 4, 8, 16, \ldots$		
c)	$5, 3, 1, -1, \ldots$		
d)	$1, 1, 2, 2, 3, 3, \ldots$		

8 The first term of an arithmetic sequence is 5 and the fourth term is 14. Find:

   a) the common difference

   b) the tenth term

   c) the sum of the first ten terms.

9 An arithmetic progression of 15 terms has first term 7 and last term –49.

   a) What is the common difference?

   b) Find the sum of the arithmetic progression.

10 The 8th term of an arithmetic progression is 9 times the 2nd term.
   The sum of the 2nd and 3rd terms is 10.

   a) Write down a pair of simultaneous equations for the first term a and the common difference d.

   b) Solve the equations to find the values of a and d.

   c) Find the sum of the first 20 terms of the progression.

     Cambridge IGCSE™ and O Level Additional Mathematics Workbook

11 A ball rolls down a slope. The distances it travels in successive seconds are 4 cm, 12 cm, 20 cm, 28 cm, etc., and are in an arithmetic progression. How many seconds elapse before it has travelled 9 metres?

12 a) How many terms of the arithmetic progression 15, 13, 11, … make a total of 55?

b) Explain why there are two possible answers to this question.

13 Are the following sequences geometric? If so, state the common ratio and calculate the eighth term.

	Sequence	Geometric? Yes / No	8th term
a)	2, 6, 18, 54, …		
b)	2, 6, 10, 14, …		
c)	1, −1, 1, −1, …		
d)	4, −12, 36, −108, …		
e)	8, 4, 2, 0, …		
f)	1, 0, 0, 0, …		

14 A geometric sequence has first term −2 and common ratio 2. The sequence has 10 terms.

a) Find the last term.

b) Find the sum of the terms in the sequence.

15 a) How many terms are there in the sequence $27, 9, 3, \ldots \frac{1}{27}$?

b) Find the sum of the terms in this sequence.

16 The 1st term of a geometric progression is positive, the 5th term is 128 and the 11th term is 524 288.

a) Find two possible values for the common ratio.

b) Find the first term.

c) Find two possible values for the sum of the first seven terms.

 Cambridge IGCSE™ and O Level Additional Mathematics Workbook

17 The first three terms of an infinite geometric sequence are 100, −60 and 36.

   a) Write down the common ratio of the progression.

   b) Find the sum of the first 10 terms.

   c) Find the sum to infinity of its terms.

18 In each month, the growth of a bush is three-quarters of the growth the previous month. The bush is initially 1.2 m tall and grows 12 cm in the first month.

   a) What is the tallest the bush will grow?

   b) After how many months is it within 5% of its maximum height?

13 Vectors in two dimensions

1 Express the following vectors **(i)** in component form and **(ii)** in column vector form.

a)
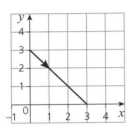

c)
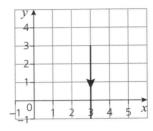

b)
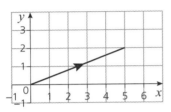

d)
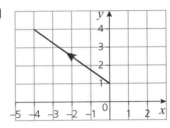

	a)	b)	c)	d)
(i) Components				
(ii) Column vectors				

2 (i) Draw the following vectors on the grid below:

    a) $-3\mathbf{i}$ **b)** $2\mathbf{i} + 5\mathbf{j}$ **c)** $4\mathbf{i} + 2\mathbf{j}$ **d)** $-4\mathbf{i} + 2\mathbf{j}$

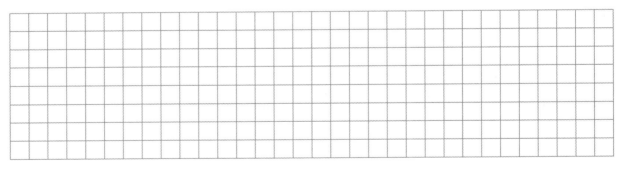

(ii) Find the modulus of each vector.

    a)   **b)**   **c)**   **d)**

3 For each of the following vectors **(i)** draw a diagram and **(ii)** find its magnitude.

a) $\begin{pmatrix} -2 \\ 0 \end{pmatrix}$

b) $\begin{pmatrix} 0 \\ 3 \end{pmatrix}$

c) $\begin{pmatrix} 5 \\ -2 \end{pmatrix}$

d) $\begin{pmatrix} -1 \\ 1 \end{pmatrix}$

4 a) A, B, C and D have coordinates $(-1, 3)$, $(1, 5)$, $(4, 3)$ and $(1, 1)$. Draw quadrilateral ABCD on the grid.

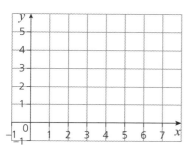

b) Write down the position vectors of the points A, B, C and D.

A

B

C

D

c) Write down the vectors:

(i) $\overrightarrow{AB}$

(ii) $\overrightarrow{BC}$

(iii) $\overrightarrow{AD}$

(iv) $\overrightarrow{DC}$

5 ABC is a triangle with $\overrightarrow{AB} = 2\mathbf{i}$ and $\overrightarrow{AC} = \mathbf{i} + 2\mathbf{j}$.

a) Write down the vector $\overrightarrow{BC}$.

c) Describe the triangle ABC.

b) Sketch the triangle ABC.

d) Find $\overrightarrow{AD}$ if ABDC is a parallelogram.

6 Simplify the following.

   a) $(3\mathbf{i} + \mathbf{j}) - (3\mathbf{i} - \mathbf{j})$

   b) $3(\mathbf{i} + \mathbf{j}) - 2(\mathbf{i} - \mathbf{j})$

   c) $2(2\mathbf{i} - 3\mathbf{j}) - 3(-2\mathbf{i} + 3\mathbf{j})$

7 $\mathbf{p} = 2\mathbf{i} - \mathbf{j}, \mathbf{q} = \mathbf{i} + 2\mathbf{j}$ and $\mathbf{r} = -\mathbf{i} + 3\mathbf{j}$

   Find the following vectors and work out their lengths:

   a) $\mathbf{p} + \mathbf{q} + \mathbf{r}$

   b) $2\mathbf{p} - 3\mathbf{q} + 4\mathbf{r}$

   c) $3(\mathbf{p} + 2\mathbf{q}) - 2(2\mathbf{p} - 3\mathbf{q})$

8 The diagram shows an isosceles trapezium OPQR where $\overrightarrow{OP} = \begin{pmatrix} 4 \\ 3 \end{pmatrix}$ and $\overrightarrow{OQ} = \begin{pmatrix} 9 \\ 3 \end{pmatrix}$.

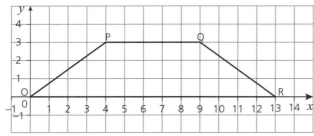

   a) Write down the vectors $\overrightarrow{PQ}$, $\overrightarrow{QR}$ and $\overrightarrow{OR}$ as column vectors.

   $\overrightarrow{PQ}$ $\qquad\qquad\qquad$ $\overrightarrow{QR}$ $\qquad\qquad\qquad$ $\overrightarrow{OR}$

   b) Write down the vector $\overrightarrow{RP}$ as a column vector.

   c) When produced, $\overrightarrow{OP}$ and $\overrightarrow{RQ}$ meet at the point B. Add B to the diagram, and use one word to describe the triangle OBR.

9 Find unit vectors parallel to each of the following:

   a) $5\mathbf{i} + 12\mathbf{j}$

   b) $\mathbf{i} + \mathbf{j}$

   c) $\begin{pmatrix} 3 \\ -5 \end{pmatrix}$

   d) $\begin{pmatrix} -2 \\ -5 \end{pmatrix}$

10 A(1, 2), B(4, 6), C(8, 9) and D(5, 5) form the vertices of a quadrilateral.

a) Draw the quadrilateral.

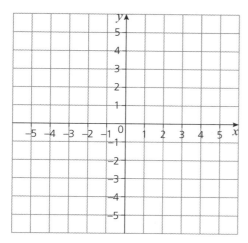

b) Write the following sides of the quadrilateral as column vectors.

(i) AB

(ii) BC

(iii) AD

(iv) DC

c) Find the lengths of the sides of the quadrilateral.

d) Describe the quadrilateral as fully as possible.

11

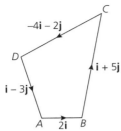

Onaka wants to measure the area and perimeter of a field that he is buying, so he asks a surveyor to measure it for him. The surveyor then presents him with this sketch.

Onaka doesn't understand how vectors work, so he starts by adding up the four pieces of information on the diagram.

a) What answer does this give him? Explain the result.

b) What answer should he get for the perimeter if each unit represents 100 m?

c) Taking A to be at the origin, write down the position vectors of the other three corners of the field.

d) Find the area of the field.

12 Jenny starts at O and travels north west for 2 hours at 6 km h^{-1}, and then east until she arrives at Q which is 20.1 km from O. Find the bearing of Q from O, giving your answer to the nearest degree.

14 Differentiation

In questions 1 to 5 differentiate the given functions with respect to x.

1 **a)** $y = x^6$ **b)** $y = 3x^2$ **c)** $y = -2x$ **d)** $y = -4$

2 **a)** $y = x^{\frac{1}{3}}$ **b)** $y = -3x^{\frac{2}{3}}$ **c)** $y = 2\sqrt[4]{x}$ **d)** $y = \frac{2\sqrt{x}}{3}$

3 **a)** $y = 4x^3 - 3x^4$ **b)** $y = 7x^2 + x - 5$ **c)** $y = 2x^3 - 3x^2 + 4$

4 **a)** $f(x) = \frac{2}{x^3}$ **b)** $f(x) = 2\sqrt{x} - x\sqrt{x}$ **c)** $f(x) = \frac{3}{2}x^{\frac{1}{2}} - \frac{1}{2}x^{-\frac{1}{2}}$

5 **a)** $y = (x - 1)(2x + 1)$ **b)** $y = 3x^2 - \frac{5}{x}$ **c)** $y = x^2(2x - 3)$

6 a) Draw the curve $y = (x + 1)(x - 2)$ **b)** Find the gradient of the curve at the points of intersection with the x and y axes.

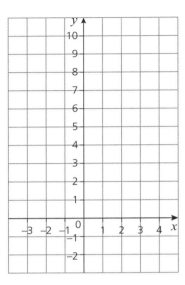

7 a) Draw the curve $y = x^2 - 4$ and the line $y = 3x$ on the same axes.

b) Use algebra to find the coordinates of the points where the two graphs intersect.

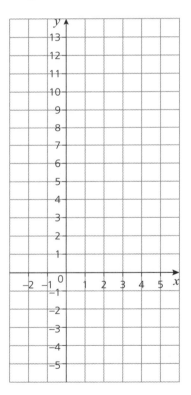

c) Find the gradient of the curve at the points of intersection.

Cambridge IGCSE™ and O Level Additional Mathematics Workbook

8 Complete the following for equations a and b.

a) $y = x^3 - 3x^2 - 9x + 15$

   (i) Find $\dfrac{dy}{dx}$ and the values of x for which $\dfrac{dy}{dx} = 0$

   (ii) Classify the points on the curve with these x-values

   (iii) Find the corresponding y-values

   (iv) Sketch the curve.

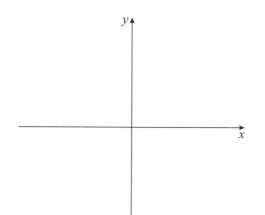

b) $y = x^4 - 8x^2 + 16$

   (i) Find $\dfrac{dy}{dx}$ and the values of x for which $\dfrac{dy}{dx} = 0$

   (ii) Classify the points on the curve with these x-values

   (iii) Find the corresponding y-values

   (iv) Sketch the curve.

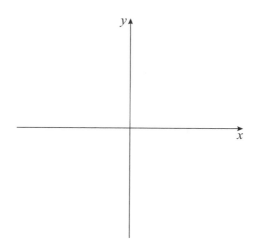

9 The graph of $y = x^2 + ax + b$ passes through the point $(-1, 10)$ and its gradient at that point is -7.

a) Find the values of a and b.

b) Find the coordinates of the stationary point of the curve.

10 a) Find the stationary points of the function $y = (x + 1)^2(x - 1)$ and classify them.

> Classifying a stationary point means determining whether it is a maximum, a minimum or a point of inflection.

b) Sketch the curve.

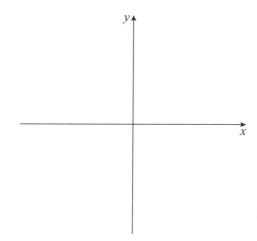

11 For each of the following curves

a) $y = 3x^3 - 4x - 4$

    (i) find $\dfrac{dy}{dx}$ and $\dfrac{d^2y}{dx^2}$

    (ii) find any stationary points

    (iii) use the second derivative test to determine their nature

    (iv) sketch the curve.

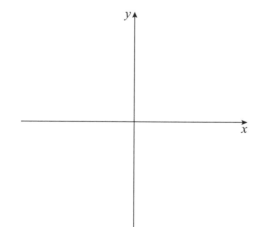

b) $y = x^4 - 6x^2 + 8x - 5$

    (i) find $\dfrac{dy}{dx}$ and $\dfrac{d^2y}{dx^2}$

    (ii) find any stationary points

    (iii) use the second derivative test to determine their nature

    (iv) sketch the curve.

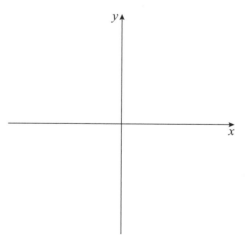

12 An open tank with a capacity of $32\,m^3$ is to be constructed with a square base and vertical sides.

a) Find an expression for the height h m of the tank in terms of the length x m of a side of the base.

b) To reduce costs, it will be constructed using the smallest possible area of sheet metal. Find its dimensions and use the second derivative test to show that your answer is a minimum.

13 A curve has equation $y = x^2 - 7x + 10$.

a) Find the gradient function $\dfrac{dy}{dx}$.

b) Find the gradient of the curve at the point P(4, −2).

c) Find the equation of the tangent at P.

d) Find the equation of the normal at P.

14 The diagram shows a sketch the curve $y = x^4 - 4x^2$.

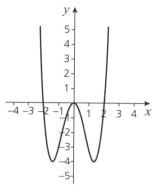

a) Differentiate $y = x^4 - 4x^2$.

b) Find the equations of the tangent and normal to the curve at the point $(2, 0)$.

c) Find the equations of the tangent and normal to the curve at the point $(-2, 0)$.

d) State the equations of the tangent and normal to the curve at the point $(0, 0)$.

15 a) The equation of a curve is $y = x^4 + x^3$. Find $\dfrac{dy}{dx}$ and $\dfrac{d^2y}{dx^2}$.

b) Find the coordinates of the stationary points on the curve.

c) Classify the stationary points, using the second derivative test where possible.

d) Sketch the curve $y = x^4 + x^3$ labelling the stationary points.

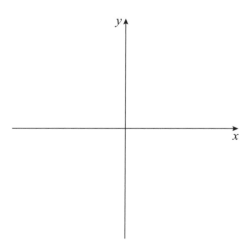

16 Differentiate each of the following functions with respect to x.

a) $y = 2\sin x + 3\cos x$

d) $y = \ln 3x$

b) $y = 2\tan x - 3\cos x$

e) $y = 4e^x$

c) $y = 3\ln x$

f) $y = 3e^{-x}$

17 Use the product rule to differentiate each of the following functions with respect to x.

a) $y = xe^x$

d) $y = x\tan x$

b) $y = xe^{-x}$

e) $y = x^2\sin x$

c) $y = x\ln x$

f) $y = x^3\cos x$

18 Use the quotient rule to differentiate each of the following functions with respect to x.

a) $y = \dfrac{e^x}{x}$

d) $y = \dfrac{e^x}{\sin x}$

b) $y = \dfrac{x}{e^x}$

e) $y = \dfrac{\ln x}{x}$

c) $y = \dfrac{\sin x}{e^x}$

f) $y = \dfrac{x}{\ln x}$

19 Use the chain rule to differentiate each of the following functions with respect to x.

a) $y = \sin^4 x$

c) $y = \sqrt{1 + 2x}$

b) $y = \tan^2 x$

d) $y = \sqrt[3]{1 + 3x}$

20 Use an appropriate method to differentiate each of the following functions with respect to x.

a) $y = e^x \cos x$

c) $y = \dfrac{\tan \theta}{1 - \cos \theta}$

b) $y = \dfrac{\ln x}{e^x}$

d) $y = (1 - \cos \theta)^2$

21 You are given that $y = \sqrt{u}$ and that $u = 2x^2 + 1$.

a) Show that the point $(2, 3)$ lies on the graph of y against x.

b) Find the values of $\dfrac{du}{dx}, \dfrac{dy}{du}$ and $\dfrac{dy}{dx}$ at the point $(2, 3)$.

15 Integration

1 Find y for each of the following gradient functions:

a) $\dfrac{dy}{dx} = 4x - 5$

b) $\dfrac{dy}{dx} = 2x^2 - 5x - 4$

c) $\dfrac{dy}{dx} = (x + 2)(2x - 3)$

2 Find f(x) for each of the following gradient functions.

a) $f'(x) = x^3 - 3$

b) $f'(x) = 4 + 3x - x^2$

c) $f'(x) = (2x + 3)^2$

3 Find the following indefinite integrals.

a) $\displaystyle\int (4x + 3)dx$

b) $\displaystyle\int (2x^4 - 1)dx$

c) $\displaystyle\int (x^3 - 2x)dx$

4 Find the following indefinite integrals.

a) $\displaystyle\int (2x - 3)^2 dx$

b) $\displaystyle\int (x + 3)(x - 2)dx$

c) $\displaystyle\int (1 - 2x)^2 dx$

5 Find the equation of the curve $y = f(x)$ that passes through the specified point for each of the following gradient functions.

a) $\dfrac{dy}{dx} = 4x + 1; (1, 3)$

b) $\dfrac{dy}{dx} = 1 - 2x^3; (4, 0)$

c) $f'(x) = (3x - 2)^2; (0, -4)$

d) $f'(x) = (x - 2)(x + 3); (-1, -2)$

6 Curve C passes through the point $(4, 10)$; its gradient at any point is given by $\dfrac{dy}{dx} = 3x^2 - 6x + 1$.

a) Find the equation of the curve C.

b) Show that the point $(2, -12)$ lies on the curve.

Questions 5–7

7 Evaluate the following definite integrals. Do not use a calculator.

a) $\int_1^3 4x\,dx$

f) $\int_{-1}^0 (5-4x)\,dx$

b) $\int_{-1}^5 6x^2\,dx$

g) $\int_0^3 (2x+1)^2\,dx$

c) $\int_{-2}^1 (x-3)\,dx$

h) $\int_{-2}^2 (2x-3)^2\,dx$

d) $\int_{-1}^2 (x^2-3x)\,dx$

i) $\int_{-1}^1 (x+1)(2x-1)\,dx$

e) $\int_{-4}^{-2} (x^3+x)\,dx$

j) $\int_1^3 x(x+1)(x+2)\,dx$

8 Find the area of each of the shaded regions. Do not use a calculator.

a)

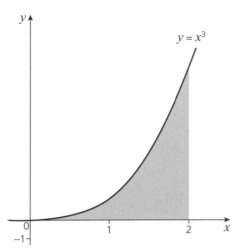

c)

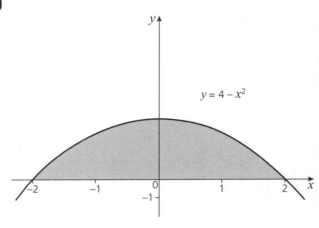

b)

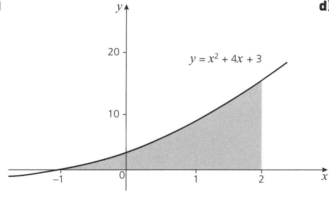

d)

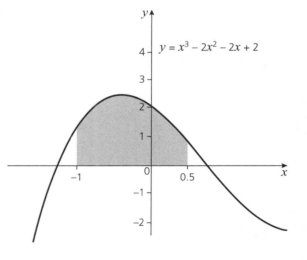

9 The graph shows the curve $y = x^2 - 4x + 3$. Calculate the area of the shaded region.

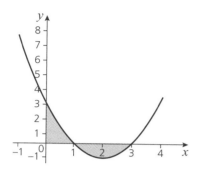

10 The graph shows the curve $y = x^3 - 5x^2 + 6x$.

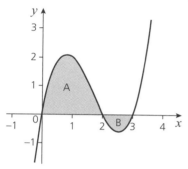

a) Find the area of each shaded region, A and B. Do not use a calculator.

b) State the total area enclosed between the curve and the x-axis for $0 \leqslant x \leqslant 3$. Do not use a calculator.

11 a) Sketch the curve $y = (x + 1)(x - 1)(x - 3)$ and shade the areas enclosed between the curve and the x-axis.

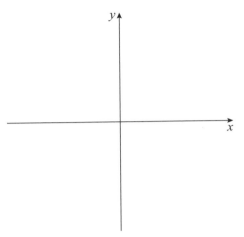

b) Find the total area enclosed between the curve and the x-axis. Do not use a calculator.

12 a) Sketch the curve $y = (x + 1)^2(x - 2)$ and shade the areas enclosed between the curve and the x-axis.

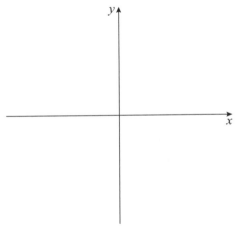

b) Find the area you have shaded. Do not use a calculator.

13 Find the following indefinite integrals.

a) $\int \dfrac{1}{2x-3}\,dx$

b) $\int e^{2x-3}\,dx$

c) $\int (2x-3)^3\,dx$

d) $\int \sin(2x-3)\,dx$

e) $\int \cos(2x-3)\,dx$

f) $\int \sec^2(2x-3)\,dx$

14 Evaluate the following definite integrals. Do not use a calculator.

a) $\int_1^4 \dfrac{2}{2x+1}\,dx$

b) $\int_1^4 e^{2x+1}\,dx$

c) $\int_1^4 (2x+1)^3\,dx$

d) $\int_0^{\frac{\pi}{2}} \sin\left(2x+\dfrac{\pi}{4}\right)\,dx$

e) $\int_0^{\frac{\pi}{2}} \cos\left(2x+\dfrac{\pi}{4}\right)\,dx$

f) $\int_{-\frac{\pi}{2}}^{\frac{\pi}{2}} \cos\left(\dfrac{x}{2}-\dfrac{\pi}{4}\right)\,dx$

16 Kinematics

1 In each of the following cases:

(i) find expressions for the velocity and acceleration at time t

(ii) find the initial position, velocity and acceleration

(iii) find the time and position when the velocity is zero.

a) $s = 3t^2 - t - 4$

(i)

(ii)

(iii)

b) $s = 4t - t^3$

(i)

(ii)

(iii)

c) $s = 5t^4 - 2t^2 + 3$

(i)

(ii)

(iii)

Photocopying prohibited *Cambridge IGCSE™ and O Level Additional Mathematics Workbook*

2 A particle is projected in a straight line from a point O. After t seconds its displacement, s metres, from O is given by $s = 3t - t^3$.

a) Find expressions for the velocity and acceleration at time t.

b) Find the time when the particle is instantaneously at rest.

c) Find the velocity when $t = 2$ and interpret your result.

d) Find the initial acceleration.

3 A ball is thrown vertically upwards and its height, h metres, above the ground after t seconds is given by $h = 2 + 10t - 5t^2$.

a) From what height is the ball projected?

b) When is the ball instantaneously at rest?

c) What is the greatest height reached by the ball?

d) After what length of time does the ball reach the ground?

e) At what speed is the ball travelling when it reaches the ground?

4 A hot air balloon is rising at a rate of $0.6\,\mathrm{m\,s^{-1}}$ and is at a height of $28\,\mathrm{m}$ when it starts to experience a downward acceleration of $0.2\,\mathrm{m\,s^{-1}}$.

a) Find the height reached by the balloon before it starts to descend.

b) How long does the balloon take to return to the ground?

c) At what speed is the balloon travelling when it reaches the ground?

5 The height of a ball thrown up in the air is given by $h = 10t - 5t^2 + 2$, where h is the height above ground level.

a) Find an expression for the velocity of the ball.

b) Find the maximum height reached by the ball and the time when this occurs.

c) Find the acceleration of the ball.

d) Find the time taken for the ball to reach the ground.

6 Find expressions for the velocity, v, and displacement, s, at time t in each of the following:

a) $a = 2 + 4t$; when $t = 0$, $v = 3$ and $s = 0$.

b) $a = 6t^2 - 2t$; when $t = 0$, $v = 6$ and $s = 2$.

c) $a = 4$; when $t = 0$, $v = 2$ and $s = 3$.

7 The acceleration of a particle $a\,\mathrm{m\,s^{-2}}$, at time t seconds is given by $a = 5 - 4t$. When $t = 0$ the particle is moving at $3\,\mathrm{m\,s^{-1}}$ in the positive direction, and is $2\,\mathrm{m}$ from the point O.

a) Find expressions for the velocity and displacement in terms of t.

b) Find when the particle is instantaneously at rest and its displacement from O at that time.

Reinforce learning and deepen understanding of the key concepts covered in the revised syllabus; an ideal course companion or homework book for use throughout the course.

>> Develop and strengthen skills and knowledge with a wealth of additional exercises that perfectly supplement the Student's Book.

>> Build confidence with extra practice for each lesson to ensure that a topic is thoroughly understood before moving on.

>> Ensure students know what to expect with rigorous practice and exam-style questions.

>> Keep track of students' work with ready-to-go write-in exercises.

>> Save time with all answers available in the Online Teacher's Guide.

Use with *Cambridge IGCSE™ and O Level Additional Mathematics*
9781510421646

For over 25 years we have been trusted by Cambridge schools around the world to provide quality support for teaching and learning. For this reason we have been selected by Cambridge Assessment International Education as an official publisher of endorsed material for their syllabuses.

This resource is endorsed by Cambridge Assessment International Education

✓ Provides learner support for the Cambridge IGCSE™ and O Level Additional Mathematics syllabuses (0606/4037) for examination from 2020

✓ Has passed Cambridge International's rigorous quality-assurance process

✓ Developed by subject experts

✓ For Cambridge schools worldwide

ISBN 978-1-5104-2165-3

HODDER EDUCATION
www.hoddereducation.com

MIX
Paper from
responsible sources
FSC™ C104740